Occupational Health [illegible]on

Muhammad Usman Ghani

Occupational Health and Safety Management System

Evaluation of Occupational Health and Safety Management System and awareness level among Employees of Textile Industry

LAP LAMBERT Academic Publishing

Impressum / Imprint
Bibliografische Information der Deutschen Nationalbibliothek: Die Deutsche
Nationalbibliothek verzeichnet diese Publikation in der Deutschen
Nationalbibliografie; detaillierte bibliografische Daten sind im Internet über
http://dnb.d-nb.de abrufbar.
Alle in diesem Buch genannten Marken und Produktnamen unterliegen
warenzeichen-, marken- oder patentrechtlichem Schutz bzw. sind
Warenzeichen oder eingetragene Warenzeichen der jeweiligen Inhaber. Die
Wiedergabe von Marken, Produktnamen, Gebrauchsnamen, Handelsnamen,
Warenbezeichnungen u.s.w. in diesem Werk berechtigt auch ohne besondere
Kennzeichnung nicht zu der Annahme, dass solche Namen im Sinne der
Warenzeichen- und Markenschutzgesetzgebung als frei zu betrachten wären
und daher von jedermann benutzt werden dürften.

Bibliographic information published by the Deutsche Nationalbibliothek: The
Deutsche Nationalbibliothek lists this publication in the Deutsche
Nationalbibliografie; detailed bibliographic data are available in the Internet
at http://dnb.d-nb.de.
Any brand names and product names mentioned in this book are subject to
trademark, brand or patent protection and are trademarks or registered
trademarks of their respective holders. The use of brand names, product
names, common names, trade names, product descriptions etc. even without
a particular marking in this work is in no way to be construed to mean that
such names may be regarded as unrestricted in respect of trademark and
brand protection legislation and could thus be used by anyone.

Coverbild / Cover image: www.ingimage.com

Verlag / Publisher:
LAP LAMBERT Academic Publishing
ist ein Imprint der / is a trademark of
OmniScriptum GmbH & Co. KG
Heinrich-Böcking-Str. 6-8, 66121 Saarbrücken, Deutschland / Germany
Email: info@lap-publishing.com

Herstellung: siehe letzte Seite /
Printed at: see last page
ISBN: 978-3-659-67704-5

Copyright © 2015 OmniScriptum GmbH & Co. KG
Alle Rechte vorbehalten. / All rights reserved. Saarbrücken 2015

DEDICATION

My work is dedicated to my beloved parents that have provided me the help in every step of my life. With their prayers I have crossed the hurdle and also achieve the difficult task of my life. I pray from ALLAH ALMIGHTY to provide him healthy and long life.

ACKNOWLEDGMENT

First of all I thank to **ALLAH ALMIGHTY** who granted me sound and peaceful mind and help me in my research work. Practical life of our Beloved **PROPHET HAZARAT MUHAMMAD (P.B.U.H)** is Ideal for me. **ALLAH** showers a lot of blessings upon him.

I have special thanks to my internship supervisor **Mr. Abrar Ul Hassan, Assistant** Director Spinning, Blessed Textile Limited, he provides me strong guidance during the period of internship.

I have also thanks to my internship supervisor, **Dr. Muhammad Waseem Mumtaz,** who provided me the kind hearted guidance in my research work.

I also thanks to my respected teachers, **Dr. Rashid Saeed, Mam Khoula Skindar, Dr. Sabiha Khurram, Mam Saira Munwar, Mam Ayesha Siddiqua, Sir Ahtasham Raza, Dr. Kiran Hina, Dr. Muhammad Imran** they provided me the help in my research work.

My research work completed with the kind hearted guidance and prayer of my beloved parents. They have supported in every time in my life. My brothers **CH. Muhammad Naveed Asim, CH. Muhammad Muneer Sarwar, CH. Muhammad Ahmad Zeshan** they have provided me the best direction in my education carrier.

I have also thanks to my friends, **Faisal Hanan, Junaid Muzffar, Muhammad Qasim, Muhammad Anees, Awais Ansar, Inam Ullah, Nasir Mehmood, Muhammad Usman Bashir, Ahsan Raza, Nabeel Asad, Fakhar Hayat, Irfan Asghar, Malik Abid, Naveed Ahmad, Kashif Sajeel,** they helped me and with their prayer, I completed my work in time.

MUHAMMAD USMAN GHANI

TABLE OF CONTENTS

CONTENTS	PAGE

LIST OF TABLES, FIGURES AND APPENDIXES

FIGURES

ABSTRACT

Occupational health and safety management system, protect the workers from the risks and hazards in the workplace. International Standard Organization (ISO) 18001 standards are applicable to health and safety and the implementation of this standard ensures the healthy work environment. The purpose of the study is to overcome the challenges, which are faced by the workers during the work. The objective of the study is to encourage the initiation towards the health, safety and well- fare of people at work, to secure the people work against hazard, to help in securing safe, hygienic work environments, to, reduce, eliminate and control hazards, to check workers are trained in terms of occupational health and safety, to create the cooperation and consultation among workers and employees and to provide the recommendation and suggestion for health and safety. The study concludes that some threats and hazards present in the workplace that creates the diseases and injuries among workers. The majority of workers works in such environment which produced diseases among them. So it is the responsibility of top management to implement the occupational health and safety management system and develops the plans to protect the workers from the hazards and also creates the awareness among them.

CHAPTER 1

INTRODUCTION

1.1 Safety

Gray, (1990) defined the situation of being free from danger of harm.

Burdine and Mc Leroy, (1992) described conditions of relative safety from unintentional injury or death due to measures designed to protect against accidents.

1.2 Occupational Health and Safety

According to world health organization WHO (1995), occupational safety and health can be defined as a multidisciplinary activity aiming at:

- Protection and support of the health of workers by eliminating work-related factors and situation hazardous to health and safety at work.

- Improvement of physical, mental and social well-being of workers and carry for the progress and protection of their working capacity, as well as specialized and social growth at work.

- Expansion and promotion of sustainable work environments and work organizations

Adeniyi, (2001) described that occupational Health (OH) is a branch of health services, especially concerned with health, safety and wellbeing of workers of all categories. It is a health service which demands that employers, both government and private should demonstrate concern for practical procedures of protecting the health of workers or employees.

1.3 Background History of Occupational Health and Safety

Entwistle, (1983) evaluated that development of current manufacturing protection movement had its roots in England, at the dawn of the 18th century industrialized revolution era. By 1750, the machinery had been invented and mining and developed industries became established. Men, women and children were employed in an occupation in factories under very poor conditions. They worked for numerous conditions and with little or no food or pure water to drink.

The development of work-related health services brought regarding the performance of safety laws and Regulations in 1833. The common certainty that accidents were bound and predictable was no longer suitable for increasing inhabitants of the English public. They argued powerfully that accidents could be controlled. People have no awareness of safety protection. They were incompetent and negligent that prohibited people don't live safely in the industrial world. They were needed the safety education and other work-related health services to protect the workers from diseases.

Barling *et al.*, (2002) evaluated that occupational health and safety practices have been given the least attention in terms of research. As a result, work-related health and safety has sustained to stay outside the normal organizational and management research.

Katsoulakos & Katsoulacos, (2007) assessed that most countries and industries hardly identify the professional health and safety practices as a vital determinant of national development. Therefore, involvement of work-related health and safety into nationwide agenda becomes an important thought for not only developed countries but also in the developing countries as well.

Meredith, (1986) described that the issues related to occupational health and safety came into consideration; there is also lack of literature on these matters. Mostly, African countries are

3

struggling with professional health and safety practices as a small number of attempts from the industries (Regional Committee for Africa Report, 2004).

The work place is a potentially hazardous environment where millions of workers gone at least one-third during their lifetime. Until 1900, this information has predictable for a long time. Since earliest times, there was a consciousness of industrial cleanliness.

In the early 20th century in the U.S., Dr. Alice Hamilton led efforts to progress industrial sanitation. She observed the industrial situation first hand and taken the information from mine owners, factory managers, and state officials with proof that there was an association between worker sickness and exposure to toxins. She also presented ultimate proposals for eliminating harmful working conditions.

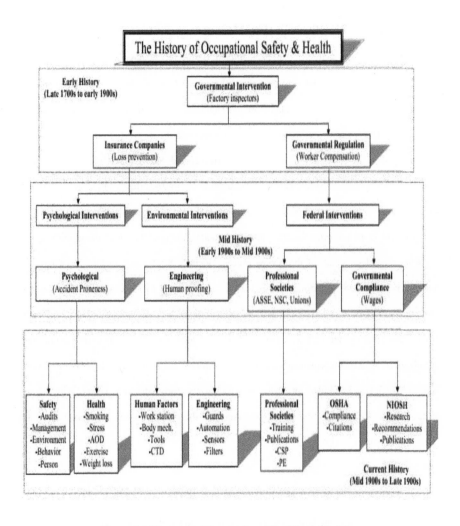

The History of Occupational Safety & Health

Early History
(Late 1700s to early 1900s)

Governmental Intervention
(Factory inspectors)

Insurance Companies
(Loss prevention)

Governmental Regulation
(Worker Compensation)

Psychological Interventions

Environmental Interventions

Federal Interventions

Mid History
(Early 1900s to Mid 1900s)

Psychological
(Accident Proneness)

Engineering
(Human proofing)

Professional
Societies
(ASSE, NSC, Unions)

Governmental
Compliance
(Wages)

Safety	Health	Human Factors	Engineering	Professional Societies	OSHA	NIOSH
-Audits	-Smoking	-Work station	-Guards	-Training	-Compliance	-Research
-Management	-Stress	-Body mech.	-Automation	-Publications	-Citations	-Recommendations
-Environment	-AOD	-Tools	-Sensors	-CSP		-Publications
-Behavior	-Exercise	-CTD	-Filters	-PE		
-Person	-Weight loss					

Current History
(Mid 1900s to Late 1900s)

Figure 1.1 Histories of Occupational Health and Safety

5

1.4 Occupational Health and Safety Management System

Abbas *et al.*, (2014) analyzed that occupational health is the physical, social and mental wellbeing in the workers of the organization at all levels and also created the encouragement and awareness about the protection among workers. The basic purpose of the International Standard Organization (ISO) 18001 implementation in any organization is to protect the workers from the hazards. The main principles of Occupational Health and Safety Assessment Specification (OSHAS) 18001 require the top management commitment for the effective function of OSHAS. To meet this criteria top management sets the objectives and targets and also done the continual improvement in its performance by conducting audits. There is also need the monitoring for checking the effectiveness of the review process, identify the hazards in the workplace, risk assessment of hazards, control the risks that are identified in the process of producing, also provide the training and creates the awareness among the workers and employees during the work about health and safety issues and also require the proper communication with stakeholders. Any of the organization can implement and adopt the OSHAS 18001. There are several benefits of the implementation of this system, risks are reduced associated with the any section of the organization, there is a good image of the organization at the market level and when the organization provides the facilities of health and safety to workers as a result cost is reduced.

1.5 Occupational Health and Safety Enforcement Practices

Boin *et al.*, (2011) described that function of the occupational Health and Safety (OSH) Act and its enforcement and collaboration in any organization, consequently called as Enforcement Act, and recognized the effect of this act on the enforcement authority of OSH. It is essential that with the passage of time the data which is collected for the research is updated according to the new rules, regulation and law, which explain that how occupational health and safety laws and legislation altered the routine of enforcement authority, how the involvement of application and practices in the system, which type of new guidance is required and it is essential to focus on the performance of OSH inspector. In this manner, it is feasible that to estimate the function of new laws and legislation also recognized the targets for the implementation process of occupational health and safety legislation. The effects of legislation can be identified through evaluation from the bottom to top or top to bottom in an organization. Top to bottom evaluation tells about the laws and decision makers' ideas and the attainment of objective of the legislation. In another way, bottom to up evaluation describes the reaction of different interested legal regulation, and also defines the goals and working on legislation. The study concludes the formation of rules and regulation about OSH, compliance and enforcement of these laws and legislation in any organization is important, and also involved the current issues to the agenda of legislation and regulatory studies.

1.6 OHSAS 18001 Standards

Abad, (2013) evaluated that the relation between the implementation of OSHAS 18001 standards and the performance of this standard in the broad point of view. First of all, implementation of OSHAS 18001 in organization done in this way to cover and identify the hazards and accidents related to the work. Second, calculate the result of OSHA 18001 standards in an association with activities related to health and safety, effectiveness of standard on labor productivity and give the importance to the feedback received from the workers about health and safety experience. The result concludes that safety variable defines the implementation of OSHAS 18001 and due to the implementation health and safety performance improved in the particular organization. The

7

experimental study concludes that there is need the valuable investment in the health and safety with planning and as a result, there is improvement in safety system and operational procedures.

The OHSAS 18001 standard is the standard for the management of occupational health and safety that deals with such system to do the audit of any organization where this system is implemented. This standard is acceptable at the international level. It describes the criteria for the organization to activate the appropriate and efficient management of occupational health and safety in an organization.

If the management system is efficiently implemented in an organization; there is identification of risks and the hazards and to reduce these risks within the premises of the organization. OSHAS ISO 18001 is proceeding in the best way, when organization checks and improves occupational health and safety system on a continual basis.

The area which covered from the OSHAS ISO 18001 Certification

- The management system is required in an organization
- Planning and identification of risk and hazard in the place of work
- Create the awareness and arrange the training for workers and employees
- Build the skills to tackle the emergency situations in place of hazards
- After the implementation of OSHAS 18001, there is continuous monitoring of workplace and also requires the improvements where gaps are present

1.7 Benefits of Occupational Health and Safety Assessment Specification (OSHAS) 18001

Desa *et al.*, (2013) described the benefits of OSHA 18001 which are as follows:

- Improve the organization image among the workers, employees, stakeholders, clients and in general public at a national and international level.
- To control the risks and hazards in the workplace, there has required the implementation of internationally accepted practices.
- At the place of work, insure the health and safety of workers, employees and stakeholder.
- Reduce the responsibilities of staff appointed under the organization.
- Create the awareness and compliance about occupational health and safety legislation
- Proper management is required to stop the accidents and incidents in the workplace
- Progress in incident examination process
- Develop such plans to build the motivation in employees
- Develop such strategies to reduce the accidents
- Show your assurance about occupational health and safety
- Implement such procedure to increase the productivity
- Decrease the risk and hazard within the limitation of laws and regulation
- Develop the better organization image by producing high quality product
- Produced the brand which provides the satisfaction to the customer
- Reduction in operational costs
- Develop good relationship among the stakeholder
- Eligibility and qualification about the business is also proved
- Implementation of this system provides the opportunity to explore the business

1.8 Occupational Health and Safety Legislation in Pakistan

Alphonse, (2008) analyzed that International Labor Organization (ILO) has used 40 different types of codes and standards which are closely link with occupational health and safety. There are two types of conventions which deal with occupational health and safety. First is occupational health and safety convention 1981 (No 155) and second is occupational health service convention 1985 (No 161) that covers the perception of occupational health and safety. There are other conventions present, which deal with the protection of workers to avoid from the risks and hazards as well as convention on health and safety specified fields of financial activities.

1.8.1 Laws in Pakistan Relating to the Issues of Occupational Health and Safety

Pakistan's Occupational Health and Safety Act are not a self-governing matter. The major law, which governs these issues, is the Chapter 3 of the Factories Act, 1934. The Hazardous occupation rule is the power of the 1963 Factories Act with additional legislation. This rule specifies some risk occupations other damaging substances, as well as the process to declare the plan as well as the top management rights.

Other relevant laws include the following:

- Dock Laborers Act, 1934

- Mines Act, 1923

- Workmen Compensation Act, 1923

- Provincial Employees Social Security Ordinance, 1965

- West Pakistan Shops and Establishments Ordinance, 1969

- Boilers and Pressure Vessels Ordinance, 2002

- Factory Act 1934

- Industrial act 1964

- Occupational health and safety act 1980

- Workplace safety and insurance act (Bill 99)

- The Fatal Accidents Act, 1855

- The Hazardous Occupations Rules, 1963

- Punjab Factories Rules, 1978

1.9 Principles of Occupational Health and Safety

ALLI, (2008) evaluated that occupational safety and health is a broad multidisciplinary field, constantly affecting on issues associated to scientific areas such as medicine – including physiology and toxicology – ergonomics, physics and chemistry, as well as technology, economics, law and other areas particular to a variety of industries and activities. In spite of this variety of concerns and wellbeing, specific principles can be familiar, including the following:

- **All workers' rights.** Workers and employees confirm that their rights are protected and they also maintain the best working condition and environment.

- More specifically:

    - The work is held in a secure and well working environment;

    - The situation of work, human self-respect and the well-being of workers must match;

    - Work should provide the different opportunities for personal achievement, self satisfaction and to serve to humanity. (ILO, 1984).

  - **Compliance of work-related health and safety policy.** These policies are implemented at the national and international level.

11

- This must be efficiently communicated to all concerned parties. Work-related health and safety, you must set up a system for the country. Such precautionary health and safety society system has everything; you require constructing and keeping mechanisms. The elements must be incorporated. Steadily build up national system and it is reviewed on the regular basis.

- National programs on work-related health and safety should be formulated. After this monitoring and assessment done on the regular basis.

- Social partners (i.e., employers and workers) and other stakeholders require the consultation. The formulation, implementation, and all the policies, systems and programs should be passed out through the evaluation.

- The basic purpose of occupational health and safety policy is to provide protection at the workplace. The main focus is done on the initial prevention. The area of working environment is identified and it is designed in such way that to create safe and healthy condition.

- Constant development of work-related health and safety must be promoted. This is necessarily that rules, regulation and laws are important tool to control the diseases, and accidents in place of work. It is best fulfilled by the growth and completion of a national policy, national system and national program.

- Particulars of the expansion and accomplishment of useful programs and policies are very significant. Precise information about the risks and damaging substances, monitoring of the work environment , policies and best practices, compliance monitoring, and other associated activities of the meeting and distribution of efficient policy making and enforcement center.

- Health support is a key part of health practice. The efforts of the employees physical, mental and social well-being should be made to get better.

- All workers should be situating to contract with the work-related health services. Preferably, all workers in all sectors of financial activity, defending and promoting the health of workers and aims to get a better working situation must be able to access these services.

- Compensation, rehabilitation and treatment services workers undergo work-related accidents, and work-related sicknesses should be able to utilize. Action to reduce the impact of industrial risks suspiciously.

- Education and exercise is a serious part of secure and healthy work environment. Workers and employers to set up safe working actions on how to perform this, you have to be familiar with the importance of this. They are specific work-related safety and health trainers to assist you solve the difficulty, for a particular industry requires special training in related fields.

- Employees, employers and the capable establishment have various responsibilities and duties. For example, workers should follow well-known safety actions; employers give a secure work atmosphere and access to emergency care, you have to check and talk with the capable establishment, and review done on regular interval and work-related health and safety policy needs to be efficient and updated.

- Policy should be applied. Work-related health and safety examination system and other measures to make sure the compliance with labor laws.

- Obviously there is overlap of the broad principles. For example, different aspects of occupational and safety and these are collected and distributed. This information based on all the areas where activities were performed. Information on the prevention of work-related accidents and diseases, as well as desired for treatment. It is also necessary that to develop such policies that are applied in the field.

1.10 Objectives of Study

❖ To promote and protect the health and safety and create the awareness among workers and employees of Blessed textile Limited.

❖ To assist in securing safe, hygienic work environment.

❖ To create the cooperation and consultation among workers and employees

❖ To identify the health and safety situation of various sections of the industry

❖ To check the workers, whether is trained in terms of occupational health and safety

❖ To provide the recommendations and suggestions for the occupational health and safety, which will give the benefit to the workers and the organization

CHAPTER 2

OVERVIEW OF THE ORGANIZATION

2.1 Introduction

This textile Industry comes under the Textile Group of Industries. This Group is an emerging field of textile and to produce the high quality products. This is located at the Faisalabad Shiekhupra Road Ferozwattoan. This is divided into the spinning and seizing sections and internship was conducted in the weaving unit of Industry.

2.2 Textile Industry (Weaving Unit)

Weaving unit is divided into following sub-section:

1- Warping section

2- Seizing section

3- Beam folding section

4- Looms section

5- Inspection and folding section

6- Weighing section

7- Packing section

8- Warehouse

9- Supply

Flow chart regarding the all sections of Textile Industry is given below.

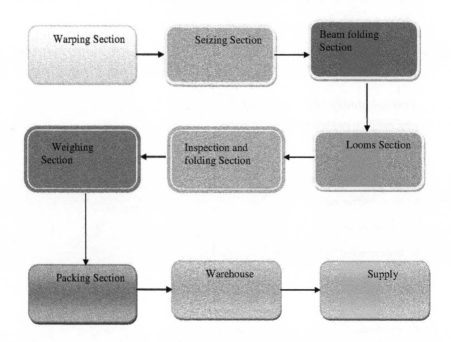

Figure 2.1 Flow chart of Textile Industry (weaving Unit)

16

2.2.1 Warping Section

Three portions are present in warping unit. In each of this portion, separate machines are working. In each section two workers are working. In warping unit, two types of machinery used. In one type of machinery, fiber is moving in an order. In this machinery different sensors are present, when fiber breaks it shows a red light. Due to which overall machinery stopped. Revolving machinery is also working on this section. The fibers are folding around beams. When fibers break, then overall procedures stops because sensors are present. Automatic switch off is also present in this area.

2.2.2 Seizing Section

It is second unit, in which beams of fiber carries from the warping unit. The workers, which carry beams with hand not use proper shoes and gloves. In seizing area, beams lifts by lifter, it is very dangerous to workers because it has not proper support. It is very high risk, due to injury or death of workers occurs. In this unit, revolving machinery is working but has not proper fencing. Emergency switch off are present. Exhaust fans and lights are present. A Chiller is not working in this area. Roof height is according to standard. In this area, fumes of starch, seizer CO and wax produce, this is inhaled by the workers because they do not use the mask, due to which they have respiratory and lung diseases. There is also very heat in this unit. Cleanliness is normal in this section. Two operators working on seizing machines, besides these four workers also work in this area. There are proper safety signs present in this area. There is not a proper emergency exit present in this section.

2.2.3 Beams Folding Section

In beam folding unit, workers work with hand. They use metal for the folding of fiber. They have not provided the Personal protective equipment (PPE). It is very high risk. Due to which workers hand's effect. In this unit, exhaust and chiller are not present. Due to which in this unit, has high heat. Workers also not use the mask, due to which they inhale dust, fumes and fluff. Workers Fire extinguisher of different types is present in this unit

2.2.4 Looms Section

In looms unit, there is the process of clothes formation. About 120 looms works, on each loom 2 workers work. Cleanliness system is better. Drier is also present, which collect the extra fluff from that area. Some fluff is collected in this way. Cleanliness system is properly managed. There are no lighting problems. In this unit, chiller system is properly working. Thermometers are present and in working condition. Temperature is maintained at 33^0C. Automatic switch-off is present. Emergency exits are present. The fire extinguisher is also present. Arrows are also present toward the emergency door. Workers are aware of with that.

2.2.5 Inspection and Folding Section

In this section, workers check the cloth on the particular stands on which glass is present. Beside the glass lights are present. Rolls of cloths are put backward on wooden stand. Machinery is also working in this place, which moves the cloth on wooden stand. A button is used for this purpose. In this unit, 30 workers works, they check the quality of cloth and any disorder present in the cloth. There is a proper exhaust system in this unit. They are underground from which they take the dust, fluff and heat. There is a proper lighting system. Workers have enough space for working. Cleanliness is best in this place. Emergency doors and fire extinguisher are present. Workers have proper training, in terms of emergency situation.

2.2.6 Weighing Section

In this unit, cloth of different type is separately come on electronic balance, where weight is checked by the workers. A lifter used in this place, for lifting the bundles of cloth. After the weighting of cloth the workers pull clothes by hand to packing section. The workers are not wearing the proper shoes and dressing, these cause different problems in workers.

2.2.7 Packing Section

In this unit, there is no high risk, because machinery is not used. Each type of cloth is separately packed in large bundles of wood. These bundles are already made, on which product put by lifter. Lighting system well managed in this area. Workers properly clean this unit. Safety boards are unavailable. When a product is packed, workers move it toward the warehouse with hands.

2.2.8 Warehouse

The warehouse is ultimately very large. Lifter is also used in this place, because heavy weight is lifted and moves to upward. Height of roof is about 16 feet. Product is placed one by one to each other to height of roof. In some places, way is present, but on the both side products are present.

2.2.9 Supply

After a specific time, the product is transported to other industry. The vehicle stands near the warehouse in the outside of weaving unit. A big slope is present near the warehouse, workers carry the product with hand to the vehicle. In this way, product transported to other industry for dying purposes.

2.4 Goals

The goal of the industry to maintain the quality of textile products at the local and international market.

2.5 Objectives

The objective of the company to maintain the ethical and professional standards at all levels.

2.6 Achievements

Blessed textile limited has following achievements

1- Machinery structure

2- Turn over

3- Social compliance

2.6.1 Machinery Structure

Machinery is working in the industry developed with high technology. Most of the machinery has an automatic switch off present. Machines are operated automatically and when there is any problem in the machinery this is automatically turned off. In warping section Benninger v creel with 1224 ends capacity each is working. In the same way, in the seizing section Benninger, pentronic-3400 is working. There are also different types of looms with a specific number of shafts working properly in the looms section. These looms have also color facility and different stability models are present. These looms are properly named as Air Jet Tsudakoma Zak e Looms. The detail all types of machinery is provided in the following table.

WIDTH		NO OF LOOMS	TYPES OF LOOMS	REMARKS
CM	INCHES			
340	134	1	R & D	SAMPLING / R & D
340	134	6	JACQUARD	STAUBULY MODEL: LX-1600
340	134	4	JACQUARD	STAUBULY MODEL: LX-1601
280	110	6	DOBBY [16 SHAFTS]	WITH FOUR COLOR FACILITY
280	110	4	POSITIVE CAM	LOOMS WITH 10 SHAFTS
340	134	18	DOBBY [16 SHAFTS]	WITH FOUR COLOR FACILITY
340	134	10	POSITIVE CAM	LOOMS WITH 10 SHAFTS
340	134	82	POSITIVE CAM	LOOMS WITH 07 SHAFTS
TOTAL		131		

OTHER MACHINES DETAILS		
WARPING	1-MACHINE	BENNINGER V CREEL WITH 1224 ENDS CAPACITY EACH.
SECTIONAL WARPER	1-MACHINE	BENNINGER,PENTRONIC-3400
SIZING	1-MACHINE	1- ZELL WITH 32 CREEL CAPACITY

Figure 2.2 air Jet Tsudakoma Zak e Looms and Machinery Details

2.6.2 Turn Over

Its annual income is about US $300 million. This group has invested more than US $25 million in current years to spread out its textile facilities. So they achieve the international demands of yarns and fabrics.

2.6.3 Social Compliance

1- Residence
2- Wages
3- Medical facility and Compensation

2.6.3.1 Residence

The workers and employees who are from the outside, they have provided with facility of residence. Workers and employees have a separate residence system.

2.6.3.2 Wages

Workers and employees have a better salary system. Salary provides at the end of the month.

2.6.3.3 Medical facility and compensation

Workers who are sick during the work they have provided the medical facility in the industry. If they have problem of injury, then the industry provides them compensation.

2.7 Products of Textile Industry

A variety of cloth is produced in the industry such as cotton cloth, warm cloth, loan cloth, wool cloth, silk cloth and curtain clothes, etc. The products are supplied within the country and exported outside the contrary. The shoe industries also use the products of cloth for the manufacturing of shoes.

2.8 Importance of ISO Certification

ISO certification is significant as it is documented worldwide as an accepted standard of quality. When companies can perfectly document their systems, they can compare them against a known standard for development. Also, because ISO certification is a familiar standard, companies can use it to measure and select the vendors or subcontractors they work with.

2.9 Industry Certifications

There are following certifications of industry:

1- ISO 9001-2008
2- OEKO -tax standard 100
3- Organic exchange standard 100 (OE-100)

2.10 Quality Policy of Industry

Textile industry committed to producing best quality fabric and ensuring on time delivering by continued improving the effectiveness of quality management system and enhancing the people skills to achieve custom satisfaction.

2.11 Hierarchy of Textile Industry

General Manager manages all the sections of textile industry under a controlled hierarchy that is appointed under this. General Manager is responsible for all the problems related to the industry. General Manager further divides its work to admin manager and mills manger. Admin manager deals with all the administrative work. Mills manger is responsible to control and manage all situations in the workplace and to report the general manger. Mills manger work is further divided into technical manager and weaving manger. Weaving manger deals all sections of weaving unit. He further reports its work to the mills manger on daily basis.

The complete hierarchy of Textile Industry is described below.

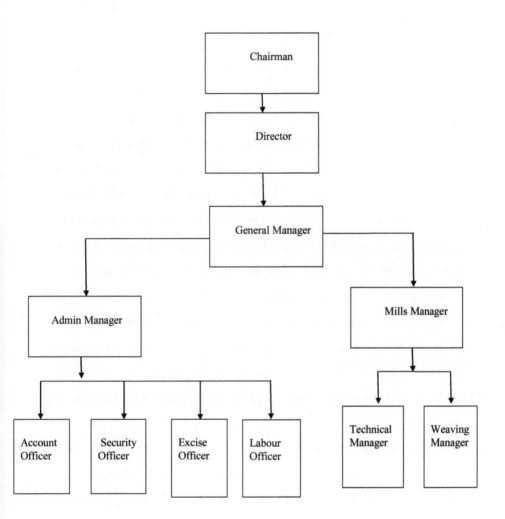

Figure 2.3 Hierarchy of Textile Industry

CHAPTER 3

REVIEW OF LITERATUE

Ahsan and Partanen, (2001) described that diseases and accidents create a hazard in the place of work. In Pakistan, occupational diseases and injuries are in high rate. The reasons for these thousands of workers are daily deals with harmful chemicals. It is accepted that healthy workers are imaginative. The use of low quality technology in the industries, results in increase accident rates, occupational diseases and dangerous working condition. Most of the workers are not aware about the protection adopted for their work because they are uneducated. Almost the whole labor force is not trying to overcome the hazards which are produced from the industrial activities. There is the lack of necessary facilities and competent person in the country that provides occupational health and safety services to the labor force. In this way the majority of workers will be at high risk if the proper planning is not done to improve occupational health and safety.

Fritschi et al., (2004) evaluated the rate of death in textile workers and studied that particular work-related contact is linked with mortality. There is a wide range of effects which are associated with textile workers such as dust, harmful chemicals and electromagnetic rays. This data was collected from present and previous members of the textile industry union from their cards of membership and National death index in which mention the date and cause of death. Total textile workers are 7684 from which death of 113 workers occurred. Both types of workers either they are male or female they have equal risk of death. But male worker's death, mostly occurs due to the injury. There is no proper information which describes a significant increase in risk link with work-related exposures.

Rana, (2005) reported that a proper system of lights is essential in stitching unit. From a safety point of view, it will avoid the workers from the eye diseases. The condition of workplace environment are not good and unclean due to which workers face different problems such as allergies, skin disease, skin rashes and also different diseases of skin. There is a lot of fluff in stitching unit which due to the aeration system spread on all the sides and cause the respiratory diseases.

Johonson & Lipscomb, (2006) evaluated that impact of large burden of work has been greater link to our health since the late 19th century. The time of work is increasing in the United States. The overall study is done on the basis of historical, sociological and work-related research and presented this as the introduction of international conference of long working hours. Research also shows that professional and educated workers work in regular hours and uneducated workers work in irregular hours. The workers, which work in irregular hours they have face the different physiological problems such as stress and chronic diseases such as cardiovascular problems. The improvement in the methodologies is required to account the exposures of long and irregular working hours. The research basically focuses on the harmful effects of stress on the health of workers. There is need to take such measures from which workers can be protected and also evaluated the stress of workers.

Saleema et al., (2007) conducted his study in a cloth mill situated in Gujarat, India. The study showed that Byssinosis generally appears in the workers, which have greater exposure during the work. The results describe that 14.42% (29) workers have the symptoms of Byssinosis. In declutching department 16 (8%) workers were affected. In Blow room 6 (3%) workers in which appear the symptoms. In a card room Byssinosis occurs in 5 (2.48%) workers. Two workers (1%) were observed in the spinning room. In total 29 workers, 16 workers belong to the age of 41-50 and symptoms appeared in those workers. This is such age in which there are greater community tasks and the age declines with the passage of time. There is the lack of occupational

27

health and safety act in the industry, due to which workers have face the different hazards and diseases such as Byssinosis and the compensation is provided to the workers.

Robson *et al.*, (2007) developed the standards, audits and guidelines of occupational health and safety management system and distributed it over 20 years. The impact of this system is timely and understandable. The aim of this study is to make the best available facts on the effects of occupational health and safety management system on employees' health and safety and linked with economic outcomes. Eight different types of bibliographic databases were used which covers the various fields of research. The overall result of the study was positive. There are some unacceptable findings, but none of them is negative findings. In spite of these good results, the study concludes that body of evidence is inadequate to formulate the recommendations in against or favor the occupational health and safety management system. Their various reasons for this, one is, different methods were used and occupational health and safety management system was studied in the original study. Second is the lack of generality of the various studies.

Makin & winder (2008) described occupational health and safety management systems (OHSMS) is an efficient system and there are some limitations which are present in the large and medium size organization, frequently in the industrial sector. There are difficulties to move the benefits of OSHMS to small scale business, with the machines and administration of the system is becoming vast. There is a need to make such framework in which insures the OHSMS is properly constructed and modified and it is presented to the individual organization and to achieve the criteria of three main control strategies that are closely link with the workplace hazards, these are safe person, safe place and safe system. The greatest attention is paid to the implementation process and builds the benefit of OHSMS more understandable. The study describes that if not all the hazards in an organization have been determined and properly addressed and also understand the process of business and also determined the factors which have a bigger impact on health and safety. There is also a minor impact of compliance auditing on the organization protection performance.

Malik *et al.*, (2010) evaluated that in Pakistan majority of workers are regularly linked with the risks and hazard in textile sector. Textile workers have exposure to the various types of hazards. Different factors are under examination that has great contribution to produce hazards in the industrial sector. There are different types of factors that are associated with the textile workers such as physical, biological, chemical and personal factors. In the working atmosphere there are different aspects that responsible to cause hazard in industrial processes such as work off shift, smoking off the place of work and negligence for the use of personal protective equipments. Technologies that create the hazard in the industry due to improper functioning of machines as a result increase in accident rate, work-related diseases and harmful working condition. Majority of workers have not awareness about to use the personal protective equipment because they are uneducated. The present study described that special measures are taken in terms of occupational health and safety to control the hazard. Random sampling method was used in which workers in the age of 30-35 and the numbers of samples were 480. The analysis of uni-variate and bi-variate represent that there is a positive and strong association. The study also shows that occupational health and safety should improve by creating the awareness about the hazards.

Hohnen & Hasle (2011) discussed the importance of certification and its impact on occupational health and safety management systems (OSHMS). Most of the research shows that this field linked to the standards such as ISO, OSHA 18001 and also deals with national and international laws and legislations. The study also describes that in cases of compliance, there is wide range of environmental issues that linked with the problem of certified organization. Moreover,

29

certification transforms the different types of topics that are addressed and various activities and procedures that are useful in an organization. A serious issue was studied in Denmark in goods manufacturing corporation for building up certified occupational health and safety system, which represent that, certified OHSMS accidently build such conditions in which facts are measured and audited. For secure working environment, these facts, not only response the markets and legal demands but also focus on the external demands for a perceptible and responsible work environmental standards. Occupational health and safety management system (OSHMS) has a greater importance and perform the function in the internal operations and external audits on global scale. In this way given the importance of the environmental matters that with the passage of time creates the guarantee for healthy environment. Certified OSHMS not necessarily deals with work environmental problems and also focus on physiological factors.

Driscol et al., (2011) assessed that the different countries have developed the list of occupational diseases to make decisions for the workers' compensation. The possible use of these lists to properly identify the diseases that link with the work that arise from appropriate exposure. The purpose of the study is to provide the broad concept that should be considered when producing a list of work-related diseases that can be established for quick determination of compensation claim. A record of occupational diseases to be used as plan for compensation purposes, this record based on a definite combination of disorder and exposure, unless a variety of exposures linked to the disorder, or most of the disorder related to specific exposure, make it unfeasible to list of all the disorder. This study concludes that there is a strong relationship between occupational exposure and disorder, there is specific criteria for identifying the disorder and the disorder consist a significant percentage of cases to identify the disorder present in all of the population and to determine in some portion of the population.

Granerud & Rocha (2011) reported a study in current decades certified management system applied to the different organization and the management of occupational health and safety by implementing the OSHA 18001 standard. An organization gets the certification, it is not necessary to meet requirement of related laws and regulation, but the most important is to improve the performance and also define goal the on the basis of health and safety continuously. The study examines that occupational health and safety management system (OSHMS) manipulate this procedure to estimate how far they delay or hold learning. This system represents such model which has the ability to examine and recognize the development processes. It was a qualitative study of Danish manufacturer which is certified with OSHA 18001 and five cases are selected at which this model is applied. The cases which are taken represent that there are greater variations in terms of OHS management in the various organizations. ISO certification has greater importance to maintain the continuous improvement to handle the health and safety issues in any organization. There is a need to improve the practices of occupational health and safety in other managerial areas. There was also increasing the employment of improvement processes within the organization. The study shows that certified health and safety management does not hinder the knowledge and can hold up higher knowledge. Improvements practices of occupational health and safety depend upon all the processes of the organization and there do not arise from the standards.

Khanzode et al., (2012) carried out a broad assessment of work-related injuries and accident's occurrence and their method of prevention. Work started from the hazard recognition, assessment of risks in the area of study, accident cause and effect and involvement strategies is discussed gradually. The things which create uniqueness in injury and accident are described. Both types of research were used such as empirical research is used to check the hypothesis and theoretical research is used for the accident origin model. Both these methods are compared and differentiated with each other. The study concludes that directions are described for work-related injury research.

Ahmad *et al.*, (2012) evaluated that in the present era, most of the industrial workers have greater exposure in industries with increasing the work-related hazards. The purpose of the study is to find out the knowledge and attitude of the workers and also determine the practice of workers in relation to occupational health. The studied was carried out in the Tribal Textile Mills, Dera Ismail Khan. The duration of the survey is about one month from 24 October 2012 to 5 November 2012. There are 650 workers in the mill from which 50 respondents are selected by sampling. Standard questionnaire was used to carry out the research. Demographic variants were selected are aged in years, different age group, language and the residence of the area. The data were analyzed by using the various statistical tests such as mean, standard deviation, percentage and frequency. T-test and One-Way ANOVA test is used to find out group wise difference of attitude, knowledge and practice of workers. And the relation among these can be determined Pearson Correlation test. The value of P is less than 0.05 that was considered statistically important. Total fifty workers, which was selected as a sample, there is no female. The Young age group is dominated and their number 44 (88%). The people which belong to the rural area are 26 (52%). The language which is dominated in that area is Seraiki with a frequency of 44 (88%) people uses this language. Two age groups of attitude and three groups of language are selected and the mean level of all these groups are different and statistically not significant. The correlation is positive and significant among the attitude, knowledge and practice of workers. In the same way effect of demographic among these three are not significant. These are concerned with occupational health and safety among workers.

32

Desa *et al.*, (2013) reported that In Malaysian occupational health and safety assessment series (OSHAS) ISO 18001 standard and occupational safety and health administration practice (OSHAP) has greater significance in an automotive factory. The purpose of the study is to recognize OSHAS 18001 standards and OSHAP and also build model of OSHAP and OHSAS 18001 in which research was carried out, these efforts was done in the Malaysian Automotive factory. The management practices of OSHA can produced decrease or increase in the different variety of measurements in the industry. This paper reviews the efforts of OSHAS 18001 and OSHAP and planned structural association model and recommendations are taken by using the structural equation modeling (SEM). Since this study is based on the previous survey, which provides the basic guidelines about this study and also develop such models which creates the relationship between the OSHAP and OSHAS ISO 18001. A hypothesis which are formulated is based on planned research and literature review. The study shows that ISO 18001 efforts which has the function to act as the mediator of the Malaysian automotive factory. This can perform such function to transform the OHS management system to Malaysian automotive industry in an efficient way. In this way, it is best among all competitors, industries of the country.

Aquil (2013) described the dynamic control of increasing occupational health and safety (OHS) engineering system at the association level and also determine the risks and hazards in the production, that are accepted by government and private zone, workers and employees. Although in Asia and Europe this standard is accepted with pleasure, but in India unexcited feedback is received as compared with environmental and quality standard. The study shows the author's experience about implementation of the safety engineering system in all manufacturing industries in India.

Floyde et al., (2013) assessed that challenge which was facing in terms of occupational health and safety at small and medium level industries. The knowledge about the occupational health and safety management and e-learning are necessary element which is used in the present research, and also done an assessment of current challenges and necessities of small level businesses. The factor which creates the barriers in the development and their appropriate solution is described in this research. Small and medium level enterprises manage and maintained the understood knowledge. Employees have to work together and cooperate in order to build this understood knowledge precise and to improve this knowledge through learning in the circumstance of the shop floor. Managers and workers show their responsibility about the awareness and the identification of aspects of occupational health and safety in place of work.

Nirmala, (2013) reported the occupational health and safety (OHS) point of view in the textile industries situated in Gujarat state of India. Developing countries are not aware of the idea of occupational health and safety and there is the limitation of knowledge about OHS in these countries. With the industrial revolution, diseases and accidents are terrible tragedy which remains constant. Therefore the occurrence of injuries and accidents are at high level. In Indian data about OHS is not properly available, the reason for this most of the accidents are not reported to the Department of Labor. Rules, regulations about occupational health and safety are not properly implemented and are poor in India. There is no modification in the laws of OHS with quickly changing the large industries with highest employee's rate in India. The sample was collected from six textile mills of Gujarat state India.

Kalejaiye (2013) evaluated that health and safety is very important and necessary for the employees and workers work in the industry and if the organization is committed to meet up its objectives and goals. In Nigeria, with the passage of time all the work is shifted toward industries

34

and machines due to which occupational health and safety problems are increasing. The problems occur as a result of exposure to dangerous chemicals, physical, mechanical, personal and biological hazards. The health problem occurred due to these hazards are less in numbers than other disabling diseases. The reason is that, workers and employees are less aware about the diseases and the pattern of sickness. The analysis of this study is done through review of literature which shows that due to hazards specific number of workers affected in Nigeria. The study examined that the pattern of compensation which is provided to workers during its duties due to the sickness, injury, and diseases in Nigeria. The study concludes that healthy workers are creative. The study gives the recommendation that health educational programs conducted in the industries to create the awareness among the employees and workers about specific hazards. In these industries, occupational health and safety management system is properly implemented and specific safety measures provided the workers to protect against hazardous, sick, injuries. The diseased workers provided the compensation against occupational hazards.

CHAPTER 4

MATERIAL AND METHODS

The study was carried out in weaving unit and respondents (workers) were considered as the target population. In this chapter basic focus on the all the data that was collected during the internship was arranged in the proper way. There was also described the methodology that was used in research work.

4.1 Research Design

The detailed descriptive study was undertaken in the Textile industry with a focus on occupational health and safety.

4.2 Methodology

The method which was used to collect the data is survey method.

4.3 Sources of Data

4.3.1 Primary Data

Primary data, regarding on the job was collected through well defined questionnaire based on chapter 3 (health and safety) of the Factory Act.

4.3.1.1 Preparation of Questionnaire

The well defined questionnaire was developed from the factory act chapter 3 that is related to health and safety. Twenty (20) questions were selected from the different sections of chapter 3. That has a main focus on occupational health and safety, diseases and awareness level among workers.

4.3.1.2 Sampling

The study was carried out in weaving units of textile industry. A number of one thousand (1000) workers work in this unit. Different sampling methods are available but the random sampling is suitable. One hundred (100) respondents were randomly selected from the weaving unit of industry with focus on the workers that spent most of time in the industry.

4.3.1.3 Primary Data Compilation

Primary data were compiled through the analysis of obtaining data. As a result percentage was shown in the form of graphs and tables. Interpretation was also given against these results. Overall results are positive and recommendations were provided for the negative results.

4.3.2 Secondary Data

The secondary data was collected from different research papers, magazines, books and newspapers published or unpublished. The data were also collected from the various websites related to occupational health and safety, environment and other related issues. The secondary data was based principally on desk analysis of literature available on the internet. This study also collected some secondary data from national and regional OHS reports and government gazettes. These literatures were searched using a series of keywords relevant to this study.

Flow chart of Methodology

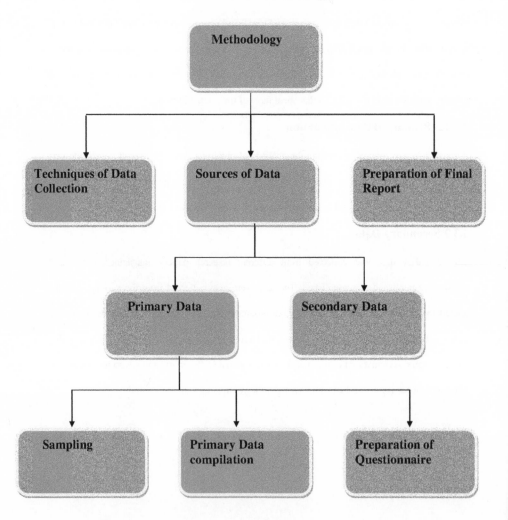

Figure 4.1 Hierarchy of Methodology

4.4 Techniques of Data Collection

For the collection of the data, properly designed questionnaire was used. Survey research method used as a collection of data. Another method of observation could not provide the satisfaction. So the survey method was best for the collection of data.

4.5 Data Presentation and Analysis

The collected data were eventually analyzed by using SPSS software. In this study, a descriptive method was used to analyze the data and information. Presentation of producing data was used in tables and necessary graphs.

4.6 Preparation of Final Report

The data which were collected either primary or secondary, used as a baseline for the preparation of the final report. On the basis of primary data, the results were shown by analyzing data through the statistical tests. Frequency test was applied to the data. As a result different charts were produced that shows the percentage of workers. The discussions were also done on the tables and graphs. Conclusion and recommendations were written on the basis of study. In other hand secondary data are helpful in the review of literature and introduction. All the data were compiled by the proper way for the submission of the final report.

CHAPTER 5

RESULTS AND DISCUSSIONS

The internship was conducted Textile industry with objective related to the evaluation and management of occupational safety and health among employees. The questionnaires were filled by the workers by asking the questions. All the data were analyzed by using the SPSS software. As results, tables and graphs were generated that shows the percentage.

Table.5.1.1 Cleanliness of walls and floor

Sr.	Categories	Percent
1	Yes	80.0
2	No	20.0
3	Total	100.0

The table 5.1.1 shows that 80% respondents said that there is the proper cleanliness, but 20% respondents said the cleanliness system is not better in some sections.

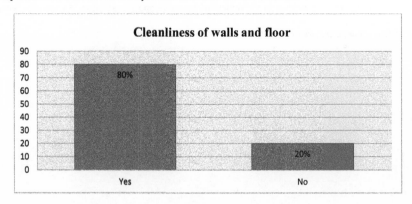

Graph 5.1.1

40

Table 5.1.2 Waste and effluents

Sr.	Categories	Percent
1	Yes	70.0
2	No	30.0
3	Total	100.0

The table 5.1.2 shows that 70% respondents said waste and effluents disposed-off properly, but 30% said disposal of waste and effluent is not proper. The majority of respondents gives their opinion that the industry has a proper waste management system for the disposal of waste.

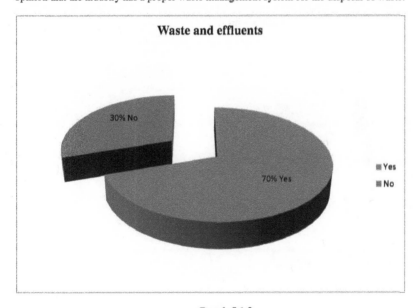

Graph 5.1.2

41

Table.5.1.3 Fumes and dusts

Sr.	Categories	Percent
1	Yes	90.0
2	No	10.0
3	Total	100.0

The table 5.1.3 shows that 90% respondents said that there is a proper system to remove the fumes during processing, but 10% said fumes are not removed from the seizing section and they complain that there are no proper equipments for the removal of fumes.

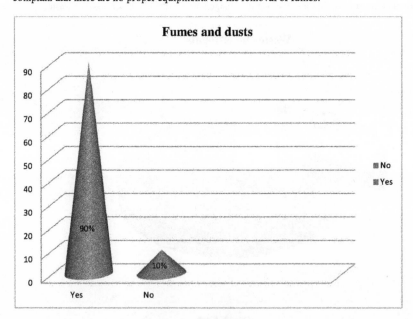

Graph 5.1.3

42

Table 5.1.4 Precautions in the confined spaces

Sr.	Categories	Percent
1	Yes	60.0
2	No	40.0
3	Total	100.0

The table 5.1.4 shows that 60% respondents said that they have provided special precautions for work in confined spaces and 40% disagree that they have not provided personal protective equipments and special precautions to work in confined spaces.

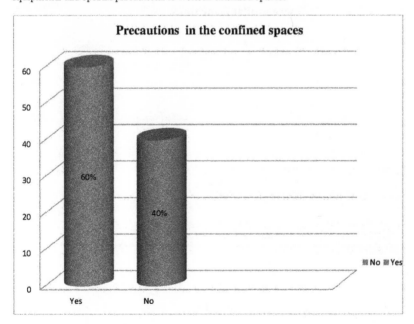

Graph 5.1.4

43

Table 5.1.5 Space available for work

Sr.	Categories	Percent
1	Yes	66.0
2	No	44.0
3	Total	100.0

The table 5.1.5 shows that 66% respondents said there have enough space to work safely, but 44% said there is not enough space for the respondents work in a safe environment and they face the difficulties during the work.

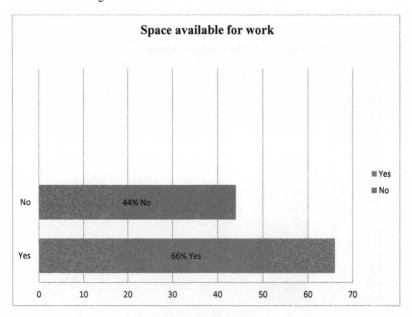

Graph 5.1.5

44

Table 5.1.6 Performance of lights

Sr.	Categories	Percent
1	Yes	73.0
2	No	27.0
3	Total	100.0

The table 5.1.6 shows that 73% respondents said there is the proper lighting to work on the machinery, but 27% said there is no proper lights present in working area as a result they face the different hazards in the form of injury.

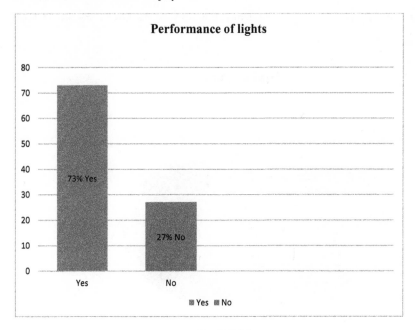

Graph 5.1.6

Table 5.1.7 Hygienic cards

Sr.	Categories	Percent
1	Yes	37.0
2	No	63.0
3	Total	100.0

The table 5.1.7 shows that 37% of respondents said they have provided the hygienic card and also necessary compensation is provided, but 63% said that they have not provided the medical facilities in the case of injuries faced during their work. In the loss of any part of body compensation does not give.

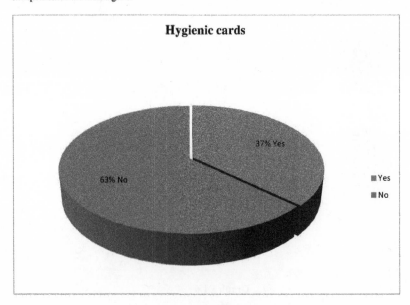

Graph 5.1.7

46

Table 5.1.8 Emergency lights and alarms

Sr.	Categories	Percent
1	Yes	45.0
2	No	55.0
3	Total	100.0

The table 5.1.8 shows that 45% respondents said that fire alarms and emergency lights properly work in all the sections of the industry and these are checked by an electrician. But 55% said that fire alarms are not in the working condition and in the case of emergency not run properly.

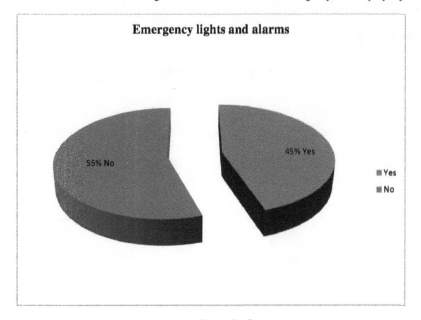

Graph 5.1.8

Table 5.1.9 Fencing around revolving machinery

Sr.	Categories	Percent
1	Yes	32.0
2	No	68.0
3	Total	100.0

The table 5.1.9 shows that 32% respondents said there is proper fencing around the revolving machinery. But 68% said that most of the machinery in the industry is revolving and there is no fence present around the machinery which leads towards the major hazards and even the loss of any part of the body during the working condition.

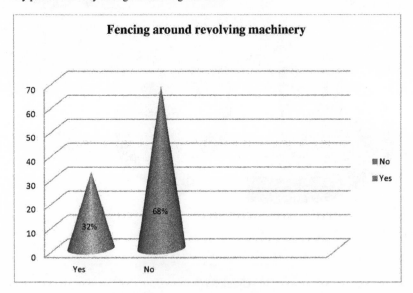

Graph 5.1.9

48

Table 5.1.10 Lung diseases

Sr.	Categories	Percent
1	Yes	93.0
2	No	7.0
3	Total	100.0

The table 5.1.10 shows that 93% respondents have lung diseases that occurred due to the fluff and dust freely moves in the industry. But 7% respondents said they have proper awareness about the use of mask and due to this reason they have not such diseases.

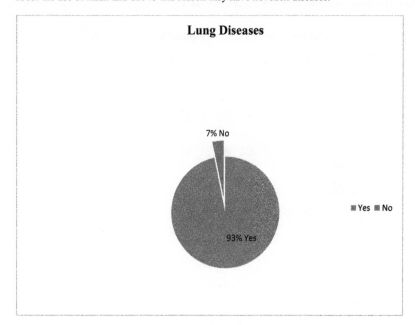

Graph 5.1.10

49

Table 5.1.11 Personal protective equipments

Sr.	Categories	Percent
1	Yes	21.0
2	No	79.0
3	Total	100.0

The table 5.1.11 shows that 21% respondents said that they have provided the masks, gloves, proper shoes and ear plug to work in processing unit and these are checked by the safety manager. But 79% said that they have not provided the personal protective equipments, due to which they have faced the variety of diseases and problems.

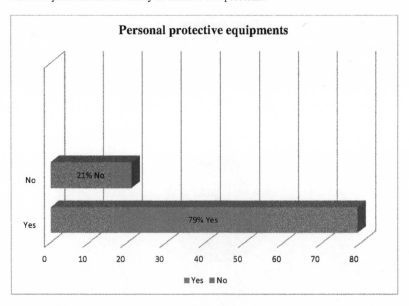

Graph 5.1.11

50

Table 5.1.12 Compensation

Sr.	Categories	Percent
1	Yes	89.0
2	No	11.0
3	Total	100.0

The table 5.1.12 shows that 89% respondents said they have given the compensation in the case of accidents or injuries if this situation occurs during the work, but 11% respondents said they have not given the proper attention by the management in case of emergency.

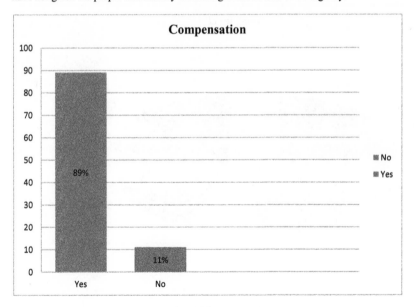

Graph 5.1.12

51

Table 5.1.13 Noise level

Sr.	Categories	Percent
1	Yes	46.0
2	No	54.0
3	Total	100.0

The table 5.1.13 shows that 46% respondents said that they are affected by the noise which is produced during the working activities and they have also the deafness problem. But 54% said they have no problem with the noise and they can work easily in this situation.

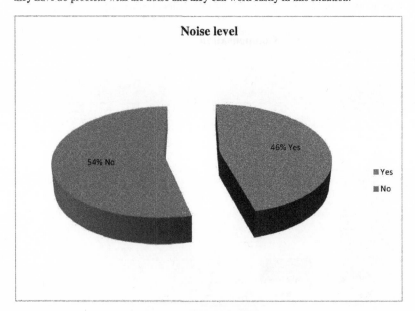

Graph 5.1.13

Table 5.1.14 Fresh and drinking water

Sr.	Categories	Percent
1	Yes	25.0
2	No	75.0
3	Total	100.0

The table 5.1.14 shows that 25% respondents said that the drinking and fresh water are available at the suitable points and they have easy access to this water. But 75% respondents said that they have no access to the freshest and drinking water and they cover a lot of distance to drink water and the proper vessels are not available.

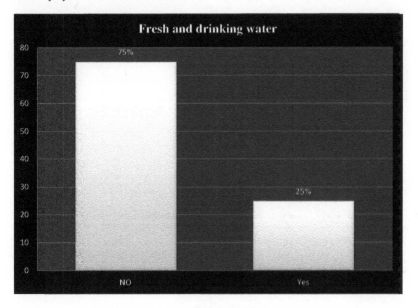

Graph 5.1.14

53

Table 5.1.15 Emergency Exits

Sr.	Categories	Percent
1	Yes	91.0
2	No	9.0
3	Total	100.0

The table 5.1.15 shows that 91% respondents said that emergency exits are present at the proper location and they have provided the proper training. In case of emergency they break these exits. But 9% said that emergency doors are not present at suitable points, in the case of fire, they cannot run properly because of the way of emergency exit different things are present.

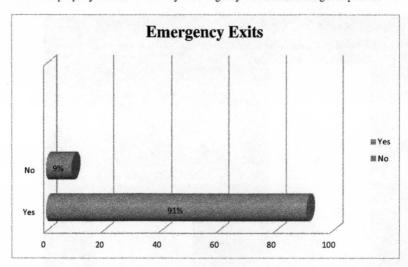

Graph 5.1.15

Table 5.1.16 Health and safety training

Sr.	Categories	Percent
1	Yes	48.0
2	No	62.0
3	Total	100.0

The table 5.1.16 shows that 48% respondents said that they have given the proper training about health and safety and awareness was created by conducting the seminars and workshop. But 62% said that they have not given the proper knowledge about the health and safety and they have also not provided the precaution in terms of safety.

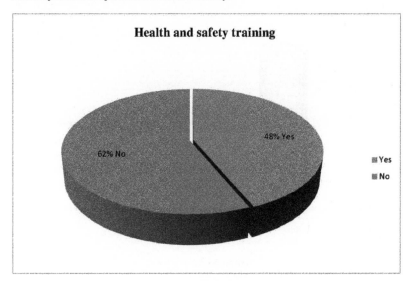

Graph 5.1.16

55

Table 5.1.17 Excessive weight

Sr.	Categories	Percent
1	Yes	12.0
2	No	88.0
3	Total	100.0

The table 5.1.17 shows that 12% respondents said they were carrying the heavy load from section to section due to which they have faced the back bone problem, but 88% respondents said they have not back bone problem because they carry the reasonable weight.

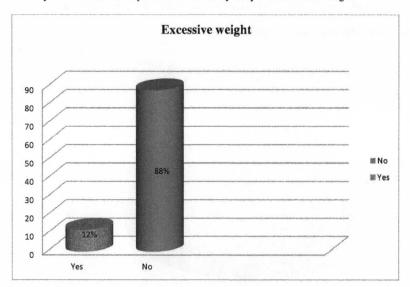

Graph 5.1.17

Table 5.1.18 Food items

Sr.	Categories	Percent
1	Yes	76.0
2	No	24.0
3	Total	100.0

The table 5.1.18 shows that 76% respondents said that the quality and quantity of food items are good. There is a proper committee, which managed all the problems related to the canteen and the prices of food items are in a reasonable range. But 24% respondents said that they are not satisfied by the quality and quantity of food items available at the canteen.

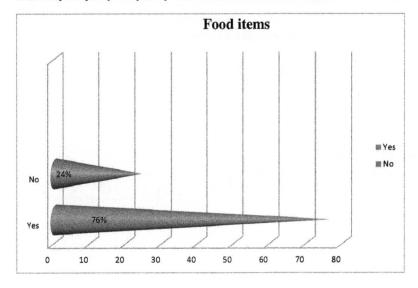

Graph 5.1.18

57

Table 5.1.19 Emergency Situation

Sr.	Categories	Percent
1	Yes	84.0
2	No	16.0
3	Total	100.0

The table 5.1.19 shows that 84% respondents said when an emergency situation occurs, then immediate action is taken by the employees and management. They use the fire extinguisher and other safety equipments according to the situation by accident. But 16% respondents said that management shows the negligence in case of emergency because most of alarms and fire extinguisher are not in working condition.

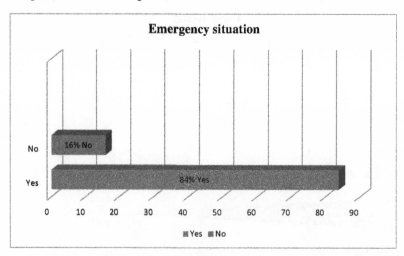

Graph 5.1.19

58

5.2 Discussions

Cleanliness is the important tool for any industry. Every organization tries their best to maintain the cleanliness in all the sections. Top management plays an important role in this. Table 5.1.1 shows that the majority of the respondents, 80% said that there is a proper system for cleanliness and this activity has done on the daily basis. But 20% respondents said that they are not satisfied with the cleanliness of the industry because in some sections, this issue has not given the proper attention.

Different types of waste are being generated in the industry. It may be solid and liquid waste. The management uses different techniques to manage the waste. The first priority of the industry is to move towards the good housekeeping. The solid waste produced is used as raw material for other industry. The table 5.1.2 shows that 70% respondents are satisfied with the proper disposal of waste and 30% said that the waste and effluents are not properly managed and disposed of.

Fumes are generated in the seizing portion of the industry. Different chemicals cause the production of fumes. These fumes moves in the seizing section and some quantity of fumes was escaping from the chimney. The table 5.1.3 shows that 90 % respondents said that they have no with the fumes and 10% said that fumes cause the suffocation and allergy problems among them. These fumes are not properly removed from the stack.

When the workers work in the confined spaces, special precautions are taken that are considered necessary for the work. Precautions may be provided by the various ways, such as creating the awareness about how to work and provided the special equipment to safe from hazard. The table 5.1.4 represents that 60% respondents said that safety manger provide the special training to work in the confined space and 40% said they have not given the precaution to work in such places.

According to the factory act, every worker has enough space to work safely in an area. Because in this way they can work with full concentration and also run the machinery in the best way. The table 5.1.5 shows that 66% respondents said they have enough space available to work safely and during the work they have not faced any difficulty and 44% said they have not provided the large space to work on the machinery or confined area as a result injuries occurred among them and they also feels the depression.

Light has also much importance in the industry, especially in the area of machines. Because the machinery is operated by hand and in any negligence injuries occur or body part may be cut. The table 5.1.6 depicted that 73% respondents said in all the sections of the industry lighting system is proper and they have not difficulties during the work and 27% said they have faced the problems due to the lights and some machinery also create the glare that may cause the accidents.

Hygienic cards are provided to the respondents when they start their work. All the detail about health and diseases is mentioned on this card. This card renewed after the six months. Workers that have diseases, haven't too done the work in the industry. The table 5.1.7 shows that 37% respondents said they have provided the hygienic card and 63% said they have not given the hygienic card.

Health and safety are the important issue in any industry. The safety may be provided to the respondents by implementing the equipments such as fire alarms and emergency lights that warn the respondents to leave the area, where the emergency occurred. This is only possible by providing the training to the respondents. The table 5.1.8 shows that 45% respondents said that they have awareness about the fire alarms and emergency lights and 55% said they have not awareness about this and these alarms and lights are out of order.

In the textile industry, most of the machinery is revolving and workers have not provided the safety instruction to work with machinery. They work in such conditions, no proper fencing present around the revolving machinery. The table 5.1.9 shows that 32% respondents said that there is fencing around the moving machinery and they have knowledge to operate the machinery and 68% said they fencing are not present around the revolving machinery and the instructions not provided to work on such machines and this may lead towards the major hazards. In the textile industry a huge quantity of fluff produced. The workers have exposure with the fluff because they do not use the masks to cover their mouth as a result, these particles move inside and effects on the lungs and also cause the respiratory diseases. The table 5.1.10 shows that 93% respondents have lung diseases and 7% said they have not exposure with fluff and they use the mask during the work.

Personal protective equipments PPE's have much importance in terms of occupational health and safety. It is the responsibility of safety manager to force the workers to use this equipment. The PPE's safe the workers from the major hazards and diseases. The table 5.1.11 depicted that 21% respondents said they have given the PPE's during the work, but no one enforce them to use these and 79% said they have not provided the PPE's as a result they face the various difficulties during the work.

Each industry has its own terms and conditions to maintain the safety situations. Health and safety may be possible by implementing ISO 18001 in the industry. Industry is responsible for the workers health and safety and in the case of any emergency or accidents, compensation provides the respondents. The table 5.1.12 shows that 89% respondents said they have provided the compensation during the work if any accidents occurred and 11% said they have not given the compensation in case of injuries or body part cut.

Noise pollution is the most important type of pollution. In looms section noise level is higher that was generated from the machines. Workers do not use the ear plug to safe from the noise. It creates the depression and deafness among workers. The 5.1.13 shows that 46% respondents said they have effected from the high noise level and it produced the deafness among them and 54% said they are not affected by the noise because they wear the ear plug and also work in the lower noise area.

Fresh and drinking water is important in the workplace. Every industry built its own water plant to provide the safe and drinking water to the workers. This water is daily changed and in the summer, cold water provides the workers. Every worker has easy access to water. The table 5.1.14 shows that 25% respondents said they have access to pure drinking water and 75% said that fresh drinking water is not providing the workers and as a result they have different water-borne diseases.

Emergency situations may occur in the industry. To tackle the situation management takes the several steps. Emergency exits are constructed in the various sections of the industry so that in any dangerous situations workers broke these exits and ran away. The table 5.1.15 shows that 91% respondents said that emergency exits are present at the proper place and they have awareness about these situations and 9% respondents said theses exits are not constructed at proper place and in most of places barriers are present in front of emergency exits.

Training and awareness are necessary components to save the workers from the hazards and diseases. Hazards are present in the workplace. To overcome these hazards top management, pay the full attention by conducting the seminars and workshops so that workers have full potential to face the hazards. The table 5.1.16 shows that 48% respondents said they have knowledge about safety situations and 62% said they have not given the proper training about the health and safety.

During the production of products workers carries the weight from the one section to another. The weight that was carried may be heavy that creates the backbone problems and injuries in the workers. The workers also do not use the proper shoes during the work that produced the hazard. The table 5.1.17 shows that 12% respondents carries the heavy weight and also have back bone problem, but 88% respondents said they do not carry the heavy weight.

Every industry has established the canteens within its own boundary. The management tries their best to maintain the quality, quantity and reasonable prices at the canteen. For this purpose they make the committee of workers to check the quality and quantity. The table 5.1.18 shows that 76% respondents are satisfied with the quality and quantity of food item, but 24% said they disagreed with quality and quantity.

The industry has developed the environmental management plan to tackle the emergency situations. This may be achieved by implementing the safety plan in all the sections and also enforce the worker to follow the guidelines regarding the health and safety. The table 5.1.19 shows that 84% respondents are satisfied that safety rules followed and in case of emergency immediate action taken, but 16% respondents are disagree with the safety system of the industry because in some section there is no proper implementation of safety rules and guidelines.

CHAPTER 6

SUMMARY/CONCLUSION AND RECOMMENDATIONS
SUMMARY

This study focuses on the evaluation and management of occupational health and safety (OHS) among employees of textile industry. Occupational health and safety is the physical, social and mental well-being of the workers at work place. International Standard Organization (ISO) 18001 is standard for Occupational Health and Safety Assessment Specification (OSHAS) that is implemented in an organization. Audit system is the necessary factor for the operation of this standard. The basic purpose of ISO 18001 standard implementation to protect workers from the issues that are related to their health and safety. For the effective management of ISO 18001, top management commitment is required. To enforce OSHAS (ISO) 18001 in the organization, different laws and legislation is currently dealing with this system. The different areas are covered after the certification of ISO 18001 such as planning and management in the identification of health related issues and monitoring is done on a continuous basis to make this system efficient. Background history of occupational health and safety start from the 17th to 19th century and brief description of workers, healthy environment, and changes in the field of occupational health and safety are discussed. Occupational health and safety legislation and those laws are present in Pakistan that deals with the workers health and safety, such as factory act 1934, workmen's compensation act 1923, industrial act 1964 and the hazardous occupation rules 1963 etc. Occupational health and safety policy must be set. This policy should be implemented and communicated to all the concerned parties. There is also the formulation of national programs in occupational health and safety. After the formulation, there is required the monitoring and assessment on regular interval. The internship carried out weaving unit of textile industry. This textile industry comes under the Umer Group of Industries, whose annual turnover

64

is US $300 million. This industry provides the basic facilities of residence and the compensation to its workers. This industry produced a high quality product. The policy of the industry is to provide the people with high quality fabric to gain the satisfaction of its customers. It was a descriptive study. The survey method was used to collect the data. Primary data are taken from the walk-through survey and secondary data was taken from the research paper, magazines and books. The data was entered and analyzes in SPSS software by applying the frequency test to the data. On average results are positive, but some are negative. By focusing on the negative result, PPE'S is major concerning issues that are not used by the workers. As a result, workers have exposure with dust, fumes and fluff which produced lung diseases in these workers. Workers have also not the access to fresh and drinking water and as a result, they have the different waterborne diseases. The safety signs and alarms are also not working in some sections. The noise level also appeared as a negative factor. It creates the deafness and depression among the workers. The study concludes that the majority of workers work in the weaving unit of blessed textile and during the work they have exposure to different types of health related issues and hazards as a result of this they have the lung problems, deafness, injuries and depression. So it is the responsibility of top management to properly build up occupational health and safety management system in blessed textile limited and also make such strategies and plans to protect and secure the workers from these health related issues.

CONCLUSION

This study analyzed the Evaluation and management of occupational safety and health among employees of textile industry. It is clear from the above discussion that occupational health and safety issues present in some sections of textile industry. The industry produced various pollutants in the manufacturing of fiber and fabrics in all the sections. During manufacturing different types of pollution generated such as water, air and noise pollution. Occupational health and safety is necessary for the organization to maintain the workers' health and to overcome the hazards which they face during the work. In all the sections health and safety conditions are satisfactory except the negligence in the use of personal protective equipments, use of safety signs and alarms, access of the workers to the fresh and drinking water and workers' awareness about occupational health and safety. There are 79% workers said that they have not provided the personal protective equipments. Due to which they have exposure to fluff and dust particles that is the cause of respiratory diseases in these workers. In the looms section noise level is high, workers aren't using the ear plug as a result they have deafness problems. On average, 75% workers have not access to the freshest and drinking water due to which they have the water-borne diseases such as hepatitis. There are 62% workers said that they have not provided the training about the occupational health and safety practices. This is negligence of top management because most of the workers are uneducated and they have not knowledge about their work as a result they have different health related issues. At the most of sections safety signs and alarms are not present, in the case of emergency workers have not practiced to tackle the situation. The workers that spend most of time in the weaving unit; they have the exposure to the dust, fluff and fumes. They have inhaled, these particles as a result, these particles move inside the body and produced the respiratory diseases. The majority of worker works in the looms area, where the

66

level of noise is high and workers work without ear plugs. After the long time exposure to noise, the problems of deafness and depressions occurred in these workers. The main objective of the study is to protect the workers from the risks and hazards produced in the workplace and also create the awareness among them. The majority of workers works in such environment which produced the problems for workers. So it is the responsibility of top management to implement the occupational health and safety management system and also develop the plans to protect the workers from the hazards and also creates the awareness among them.

RECOMMENDATIONS

- Use the personal protective equipment (PPE'S) such as shoes, ear plug, mask and protective clothing etc. to protect from the risks, hazards and injuries.

- Use the scrubbers to control dust and fluff in the seizing and looms section.

- Use fire alarms and emergency lights at specific location so that it is easily hear and seen by the workers during its work.

- Use the different type of fire extinguisher in such place where is the danger of fire and workers have the knowledge about their use and after a specific time period, these are refilled.

- Provide the training to the workers and employees to overcome the situations of emergency.

- Use the appropriate thermometer in all sections of weaving unit to check that the temperature is within the range.

- For higher temperature uses the chillers and exhaust so that the intensity of temperature is decreased.

- Implement the environmental management system in the industry to reduce the risks and hazards in the workplace.

- To control the noise level uses, such machinery which works in an efficient way and also provides the lubricant to the parts of machine decrease the noise level.

- There is also another tool to control the noise is to build the sound proof rooms so that noise that is created in the particular section not disturbs the other people.

- To control the vibrations of machines increase the area where the machines are running and also used the best construction material that creates less vibration in the earth.

- To identify the risks and hazards in the workplace and also provide the environmental management plan against these hazards.

- To display the safety signs and boards in all the places where hazards and injuries may occur and these are written in Urdu or present in graphical form.

- Chemicals and other dangerous material present in the safe place and the safety precaution mention about their usage and storage.

- Proper fencing is present around the revolving machinery to protect the workers from injuries.

- To improve the health status of workers provides the freshest and drinking water to the workers in a suitable location near to the work place.

- To maintain the quality of food items in the canteen, there should be a team nominated by the workers, which check the quality, quantity, prices and cleanliness in the canteen.

- Air pollutions which are produced during the industrial activities are controlled by putting the scrubbers and filter in the stack so that all particulates not emit into the atmosphere.

- A solid waste that was generated, it is first of all segregated and useful material use of the raw material and other material is dumped at the suitable location.

- To create the awareness among workers and employees there is necessary to conduct the workshops and seminars and also provide the training to the workers about health and safety.

- Top management enforces the workers to follow the occupational health and safety practices to overcome the health related issues.

REFERENCES

- Adeniyi JA (2001), Occupational health: A fundamental approach Haytee Organization. 46(5).

- Alphonse, T. (2008), Decent work and global strategy on occupational safety and health. On Guard Sustainable Safety and Health Practice. Vol 15 (1): 7-13.

- Abad *et al.*, (2013), An assessment of the OHSAS 18001 certification process: Objective drivers and consequences on safety performance and labour productivity, Journal of Safety Science ,Volume 60, Pages 47–56.

- Ahasan, M.R. and T. Partanen, (2001), Occupational health and safety in the least developed countries– a simple case of neglect. J. Epidemiol. 11(2): 74-80. Malik et al. (2010), Role of hazard control measures in occupational health and safety in the textile the industry of Pakistan., Pak. J. Agri. Sci., Vol. 47(1), 72-76;

- Ahmad *et al.*, (2012), Knowledge, attitude and practice related to occupational health and safety among textile mills workers in Dera Ismail Khan, Gomal J Med Sci; 10: 222-6.

- A.M. Makin C. Winder, (2008), A new conceptual framework to improve the application of occupational health and safety management systems, Safety Science Volume 46, Issue 6, Pages 935–948.

- Aghera Nirmala D, (2013), A Study of Workplace Hazards Faced by Workers in Textile Industry as a Part of Occupational Health and Safety, Paripex- Indian journal of research, Volume : 2, page 333-334.

- Burdine JN, McLeroy KR (1992), Practitioners' use of theory: Example of safety education. Health Educ. Quart. 19(3).

- Boin et al., (2011), a time for public administration. Public Administration 89 (2), 221–225.

- Barling *et al.*, (2002), Development and test of a model linking safety-specific transformational leadership and occupational safety, Journal of Applied Psychology, 87, 488– 496.

- Benjamin O. ALLI (2008) Fundamental principles of occupational health and safety, International Labour Office – Geneva: second Edition, Page 17-19.

70

- Desa *et al.,* (2013), The Impact of Occupational Safety and Health Administration Practices (OSHAP) and OHSAS 18001 efforts in Malaysian Automotive Industry, Journal of Applied Science And Research, , 1 (1):47-59.

- Entwistle IR (1983), Adventures in industries and aviation, British Medical Journal, London. Kalejaiye, (2013), Occupational health and safety: Issues, challenges and compensation in Nigeria, Peak Journal of Public Health and Management Vol.1 (2), pp. 16-23.

- Fritschi *et al.,* (2004), Mortality in Australian Cohort of Textile workers. Journal of Occupational Medicine, 54(4): 225-257.

- Floyd *et al.,* (2013), the design and implementation of knowledge management systems and e-learning for improved occupational health and safety in small to medium sized enterprises, Journal of Safety Science Volume 60, Pages 69–76.

- Gray JE, (1990), Planning health promotion at the worksite Indianapolis: Benchmark press.

- Jeffrey V. Johnson, and Jane Lipscomb, (2006), long Working Hours, Occupational Health and the Changing Nature of Work Organization, American journal of industrial medicine 49:921–929.

- Katsoulakos, T., & Katsoulacos, Y. (2007), Integrating corporate responsibility principles and stakeholder approaches into mainstream strategy: A stakeholder-oriented and integrative strategic management framework. Corporate Governance, 7(4) 355–69.

- Meredith, T. (1986), Workers' health in Africa. Review of African Political Economy, 13(36), 24 – 29. Regional Committee for Africa Report. (2004). Occupational health and Safety in the African Region; Situational Analysis and perspectives. Fifty-fourth Session (WHO) Brazzaville, Republic of Congo, Africa, 1-25.

- Muhammad Aquil, (2013), Safety Engineering Practices in Processing Industries: An Analytical Survey, Journal of global research analysis, Volume : 2 page 78-79

- Malik et al., (2010), Role of hazard control measures in occupational health and safety in the textile the industry of Pakistan , Pak. J. Agri. Sci., Vol. 47(1), 72-76;

- Pernille Hohnen, Peter Hasle, (2011), Making work environment auditable – A 'critical case' study of certified occupational health and safety management systems in Denmark Safety Science Volume 49, Issue 7, Pages 1022–1029.

- Peter O. KALEJAIYE, (2013), Occupational health and safety: Issues, challenges and compensation in Nigeria, Peak Journal of Public Health and Management Vol.1 (2), pp. 16-23.

- Rana, I.M. 2005. Work places in industries. The daily Dawn. p.18. Malik et al. (2010),), Role of hazard control measures in occupational health and safety in the textile the industry of Pakistan, Pak. J. Agri. Sci., Vol. 47(1), 72-76.

- Researcher Lise Granerud, Robson Sø Rocha (2011), Organisational learning and continuous improvement of health and safety in certified manufacturers, Journal of Safety Science, Volume 49, Pages 1030–1039.

- Robson et al., (2007), The effectiveness of occupational health and safety management system interventions: A systematic review, Safety Science, , page 329–353 vol. 45.

- Saleema et al., (2007), Determination of cotton dust concentration in different textile mill in Gujarat and Prognostic evaluation of Byssinosis in its workers.www.iepkc.org (retrieved by on March 13, 2008).

- Tim Driscoll, Mark Wagstaffe and Neil Pearce, (2011), Developing a List of Compensable Occupational Diseases: Principles and Issues, the Open Occupational Health & Safety Journal, 3, 65-72.

- Vivek V. Khanzode, J. Maiti, P.K. Ray (2012), Occupational injury and accident research: A comprehensive review, Journal of Safety Science Volume 50, Pages 1355–1367.

- Zain Abbas , Muhammad Qasim and Aroj Bashir , (2014), Occupational health and safety and environmental conditions at Faruki Pulp Mills Pvt Ltd. International Journal of Science, Environment and Technology, Vol. 3, No 2, 561 – 570.